The INSIDE GUIDE

EARTH'S EXTREME PLACES

Rainforests

By Jennifer Lombardo

Cavendish Square

Published in 2026 by Cavendish Square Publishing, LLC
2544 Clinton Street Buffalo, NY 14224

First Edition

Library of Congress Cataloging-in-Publication Data

Names: Lombardo, Jennifer author
Title: Rainforests / Jennifer Lombardo.
Other titles: Rain forests
Description: Buffalo, NY : Cavendish Square Publishing, [2025] | Series: The inside guide: Earth's extreme places | Includes bibliographical references and index. | Audience term: juvenile
Identifiers: LCCN 2024061892 | ISBN 9781502674500 library binding | ISBN 9781502674494 paperback | ISBN 9781502674517 ebook
Subjects: LCSH: Rain forests–Juvenile literature | Rain forest ecology–Juvenile literature
Classification: LCC QH541.5.R27 L66 2025 | DDC 577.34–dc23/eng/20250218
LC record available at https://lccn.loc.gov/2024061892

Editor: Jennifer Lombardo
Copyeditor: Nicole Horning
Designer: Deanna Lepovich

The photographs in this book are used by permission and through the courtesy of: Cover Tahseenamjad/Shutterstock.com; p. 4 Jhampier Giron M/Shutterstock.com; p. 6 grayjay/Shutterstock.com; p. 8 Zack Frank/Shutterstock.com; p. 9 Eray Bozkurt/Shutterstock.com; p. 10 ieppei/Shutterstock.com; pp. 12 (top), 28 (bottom) Henk Bogaard/Shutterstock.com; p. 12 (bottom main) Andrey Novgorodtsev/Shutterstock.com; p. 12 (bottom inset) mr_tigga/Shutterstock.com; p. 14 Dr Morley Read/Shutterstock.com; p. 15 (inset) Michail_Vorobyev/Shutterstock.com; p. 15 (main) Hlib Shabashnyi/Shutterstock.com; p. 16 PARALAXIS/Shutterstock.com; p. 18 (main) jasmal - jasmal/Shutterstock.com; p. 18 (inset) Daney daney/Shutterstock.com; p. 20 Richard Whitcombe/Shutterstock.com; p. 21 Marcio I. Sa/Shutterstock.com; p. 22 Hans Elmo/Shutterstock.com; p. 24 AnaLysiSStudiO/Shutterstock.com; p. 25 SL-Photography/Shutterstock.com; p. 27 ChameleonsEye/Shutterstock.com; p. 28 (top) laponpat maliwan/Shutterstock.com; p. 29 (top) Andres_Rivas/Shutterstock.com; p. 29 (bottom) Paulo Jr/Shutterstock.com.

CPSIA compliance information: Batch #CSCSQ26: For further information contact Cavendish Square Publishing LLC at 1-877-980-4450.

Printed in the United States of America

CONTENTS

The Amazon is the largest and best-known rainforest in the world.

TROPICAL AND TEMPERATE

Chapter One

A rainforest is, as its name suggests, a forest that receives a lot of rain. The two main types of rainforests are tropical and temperate. Tropical rainforests cover the most land and are the best-known. To be considered a tropical rainforest, the area must get at least 80 inches (203 cm) of rain per year. All this rain makes it easy for plants and animals to **thrive**.

Tropical

Tropical rainforests are the type most people picture when they hear the word "rainforest." Tropical rainforests are mainly found near the equator. There are tropical rainforests in Africa, South America, Southeast Asia, and Australia. Altogether, tropical rainforests cover about 7 percent of the planet's surface.

The largest rainforest in the world is the Amazon, and it contains the world's largest river, which is also called the Amazon. Most of the Amazon Rainforest is located in

Fast Fact

The Amazon is three times larger than the second-biggest rainforest, which is called the Congo Basin.

This map shows the locations of the world's tropical rainforests.

Brazil. However, it also covers parts of Peru, Colombia, Venezuela, Bolivia, Guyana, Ecuador, Suriname, and French Guiana.

Fast Fact

French Guiana is a South American area that is controlled by France. It is different than the country of Guyana.

Tropical rainforests are warm and wet all year round. The temperature ranges from 68 degrees Fahrenheit (20 degrees Celsius) to 84 degrees Fahrenheit (29 degrees Celsius). This climate is most consistent near the equator. At the northern and southern edges of the rainforest, temperatures and rainfall vary more widely. Some parts of tropical rainforests even have dry seasons and rainy seasons. The rainy seasons are sometimes called monsoons.

OTHER RAINFOREST TYPES

Although the two main types of rainforests are tropical and temperate, there are **subcategories** of rainforests, including:

- Lowland: This type of rainforest is found below 3,280 feet (1,000 m) in tropical areas.
- Montane: Montane rainforests grow in mountain regions. Lower montane forests are found between 3,280 and 8,200 feet (1,000 and 2,500 m), while upper montane forests grow between 8,200 and 9,842 feet (2,500 and 3,000 m). The upper montane is sometimes called a "cloud forest" because it is often covered in fog.
- Mangrove: This type of rainforest is full of plants that grow well in **brackish** water and loose soil.

Fast Fact

Temperate rainforests contain some of the slowest-growing and oldest-lived plants in the world.

Temperate

The other main kind of rainforest is a temperate rainforest. This type of rainforest can be found along the coast in places such as western North America, western Chile and Argentina, and southern Japan. The temperature in temperate rainforests has a much lower range than tropical rainforests—from 39°F (4°C) to 68°F (20°C).

Temperate rainforests also get less rain than tropical rainforests, but still more than other parts of the world. To be categorized as a

Visitors to Olympic National Park in Washington State can hike through the temperate Hoh Rainforest, shown here.

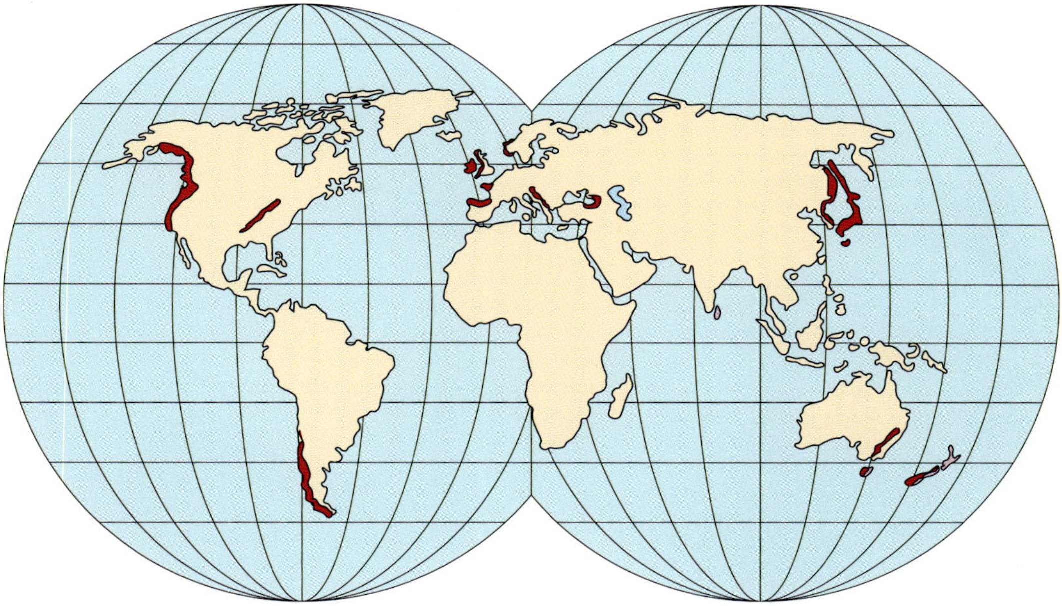

This map shows the location of some of the largest temperate rainforests in the world.

temperate rainforest, an area must get at least 55 inches (140 cm) of precipitation per year. This type of rainforest is not as **biodiverse** as tropical rainforests, but it does have many plants and animals that are not found in any other type of climate.

Another difference between tropical and temperate rainforests is that the climate in tropical rainforests is generally the same all year round, while temperate rainforests have four distinct seasons. For this reason, some of the precipitation in a temperate rainforest comes from snow.

Many different kinds of animals live side by side in tropical rainforests. Shown here are a marmoset (*left*) and a sloth (*right*).

BIODIVERSITY HOTSPOTS

Chapter Two

Tropical rainforests are some of the most biodiverse places on the planet. In fact, they are home to more than half of all the species, or types, of plants and animals in the world.

Scientists organize tropical rainforests into layers. Different plants and animals can be found in each layer, but some animals move between the layers.

Forest Floor

The bottom layer of the rainforest is the floor, or ground. If the rainforest has a river in it, such as the Amazon, it is included in this layer. The forest floor is dark and wet because very little sunlight reaches it.

The ground in a rainforest is generally covered in litter. In a city, "litter" **describes** items that belong in a trash can, such as candy wrappers, tissues, or chewing gum. However, in a rainforest, "litter" describes a layer of rotting leaves and the bodies of animals that have died. As the litter breaks down, it **nourishes** the soil. This nutrient-rich soil, combined with

Fast Fact

Scientists believe tropical rainforests first formed around 60 million years ago. This makes them one of the oldest ecosystems in the world.

Poison frogs are tiny and brightly colored. They live on the floor of rainforests in places such as the Amazon and the country of Costa Rica.

the heavy rainfall, is what makes plants thrive in a rainforest.

The animals that spend most or all of their time on rainforest floors around the world include big cats, such as tigers and jaguars;

The piranha is a small fish with very sharp teeth. It lives in the Amazon River as well as other South American rivers. Piranhas hunt in groups, which allows them to take down larger prey than other fish their size. Many people are afraid of piranhas, but they almost never attack humans.

TEMPERATE RAINFOREST WILDLIFE

The plants and animals in temperate rainforests are less diverse than those in tropical rainforests. While a tropical rainforest contains thousands of species that are found nowhere else in the world, temperate rainforests generally have fewer than 50 **unique** species. In North America, animals such as black bears, woodpeckers, elk, and mountain lions spend time in temperate rainforests as well as in other habitats.

elephants; gorillas; turtles; crocodiles; fish; warthogs; and cassowaries. Different animals are found in different rainforests. For example, the cassowary—a large, flightless bird similar to an ostrich—lives in the tropical rainforests of Australia and New Guinea.

Fast Fact

Every rainforest, except for those in Australia, has its own species of cat. For example, tigers live in India, jaguars live in the Amazon, and leopards live in Africa and Asia.

Understory

The layer above the forest floor is called the understory. This layer includes the trunks and lower branches of tall trees, hanging vines, and short bushes and trees. The animals that spend most of their time in the understory are good climbers.

Many different kinds of bugs—such as ants, butterflies, beetles, and mosquitoes—can be found in the understory. Larger animals such as

Ferns, moss, and vines grow well in the understory.

Fast Fact

Snakes such as boa constrictors and big cats such as jaguars often move between the forest floor and the understory.

monkeys, lizards, bats, and possums also spend much of their time here.

Because the understory is higher up, it gets a little more light than the forest floor, but not much more. The plants that grow here have **adapted** to their environment, or surroundings. One example is the pitcher plant. This is a carnivorous, or meat-eating, plant that gets extra nutrients by trapping bugs and small animals. The plant makes special chemicals that **dissolve** the animals over time.

Canopy and Emergent

The third layer is the canopy. It gets its name because it is made up of the tops of tall trees, which overlap to create a **dense** canopy, or covering, over the other layers of the rainforest. This is why the lower layers are so dark: The trees that make up the canopy block most of the sunlight.

The animals that live in the canopy spend most or all of their time in the trees. This is where most rainforest birds can be found, as well

The difference between a monkey and an ape is that monkeys have a tail, and apes do not. Gorillas are large, heavy apes that spend most of their time on the ground, although they do sometimes climb trees for fun or to find food. Orangutans (*shown here*) are lighter apes that spend almost all of their time in the canopy.

as animals that are good climbers, such as monkeys, sloths, and lemurs.

The last layer of the rainforest is called the emergent layer. This is where some trees emerge, or rise above, the canopy. Because the emergent layer gets the most sunlight, it is the driest layer of the rainforest. Some animals that spend time in the canopy can also be found in the emergent layer, such as birds, sugar gliders, and some of the smaller species of monkeys. The branches of emergent layer trees are thinner than canopy trees, so they can't support heavy animals.

This picture shows a tree in the emergent layer. Unlike in the canopy, where the trees all touch, the trees in the emergent layer are spaced farther apart. For this reason, most of the animals that live in this layer fly or glide.

Açai is a nutrient-rich berry that grows in the Amazon Rainforest. This picture shows a boat full of a berry harvest.

Chapter Three

THE LUNGS OF THE PLANET

Rainforests, especially tropical ones, are an important part of life on Earth. They are often called "the lungs of the planet" because there are so many plants in each rainforest. These plants absorb carbon dioxide from the air and release oxygen for people and animals to breathe.

Rainforests also store much of Earth's water and release it back into the air. If they did not do this, there would be many more **droughts**. Droughts can cause crops to die, leading to widespread hunger and disease. Droughts also increase the risk of wildfires.

Important Plants

It is hard to express how important rainforest plants are to the survival of the planet and the creatures that live on it—including humans. In addition to cleaning the air, rainforest plants are important sources of food, medicine, **commercial** products, and more. In fact, rainforests play a much bigger role in daily life than most people realize.

> **Fast Fact**
> Too much carbon dioxide in the air causes global warming, which leads to climate change.

Cinnamon comes from the bark of a tree that is native to the rainforests of Southeast Asia. The bark curls up as it dries, and it is sold as whole sticks and in powdered form.

Many things people around the world eat, drink, or use every day come from plants that grow in a tropical rainforest. Some examples include:

- chocolate
- coffee
- cashews
- cinnamon

Fast Fact

Experts estimate that about 25 percent of all the world's medicines come from plants, and 70 percent of plants that can help fight cancer grow only in rainforests.

NATURAL AND SYNTHETIC

In most cases, a specific plant is only a medical treatment because of one chemical within that plant. When scientists are able to figure out what that is, they can come up with ways to make it synthetically, or in a lab. This takes time and effort, but it is worth it because making ingredients synthetically gives the world a larger supply without further endangering rainforest plants.

One example is called diosgenin. The only place it is found naturally is in wild yams from Mexico and Guatemala. Diosgenin is one of the active ingredients in birth control medications. For many years, scientists took the diosgenin out of the yams. When they finally figured out how to make it synthetically, they no longer needed to harvest so many wild yams, saving many rainforest plants.

- Brazil nuts (eaten as well as used in moisturizers)
- quinine (a medicine used to treat a disease called malaria)
- cancer treatments
- mahogany (a wood used for furniture)

One important rainforest plant is the rubber tree. People use the sap from this tree to make rubber products, including rubber bands, erasers, car tires, gloves, balloons, boots, medical supplies, and thousands more. The trees can be grown on large plots of land called plantations, but other trees need to be cut down to make room for them. When too many trees are cut down, it is called deforestation.

This picture shows deforestation in action.

Rainforests in Danger

Deforestation is one of the biggest problems facing rainforests around the world today. People cut down trees to make room for plantations of a single crop that makes a certain product, such as rubber, palm oil, or coffee. This is because there is a lot of money to be made in selling those products. They also cut down trees to sell for their wood, clear land to raise sheep and cows, build roads, and mine for gold and other metals. The act of clearing the land also pollutes, or dirties, the air.

This problem has gotten worse in recent years. Deforestation is happening faster as companies grow and try to meet increasing demand. In 2023, a study published by a group that works to protect the Amazon found that in the previous 5 years, the rainforest lost enough trees to almost completely cover the country of Great Britain.

Deforestation causes many problems. For example, it makes climate change worse. With fewer trees to absorb carbon dioxide from the air and release oxygen, the planet will have less clean air and increased global warming. Since rainforests also affect the water cycle, deforestation also means there will be less water, and the water that is available will be more polluted.

The biggest action consumers, or shoppers, can take to protect the rainforests is to be careful about what they buy. They can read the list of ingredients on food, shampoo, soap, and other items, and avoid any that list palm oil, because palm oil plantations are one of the biggest causes of deforestation. They can check whether the label on the products they are buying says they have been made sustainably, or in a way that protects the environment. Cutting back on eating meat also creates less demand for land to raise cattle. You can write to your elected officials and ask them to get involved in international organizations that protect rainforests. If we all work together, we can make a difference.

Fast Fact

Between 2001 and 2020, the Amazon lost 9 percent of its forest cover due to human activity.

Clearing land for cattle to graze, or eat, on causes multiple problems. Deforestation is one. Another is that cows release a gas called methane when they burp. The more cows there are, the more methane is released into the air, and methane traps more heat in the air than carbon dioxide.

People who live in the rainforest generally do not have electricity. Shown here is a traditional way of cooking a meal.

Chapter Four

RAINFOREST LIFE

About 50 million people live in tropical rainforests around the world. Many are Indigenous, or Native, people whose families have lived in rainforest areas for thousands of years. Today, there are still some "uncontacted" groups living in rainforests, especially the Amazon. With so much land filled with dense forest, it is easy for people who know the area well to hide from outsiders.

Good and Bad

Like everywhere else on Earth, life in a rainforest has its ups and downs. People who have lived in these areas say the heat is one of the most noticeable things. A tropical rainforest is always hot, and there is little to no breeze, especially in the bottom layers of the forest. People can escape the heat for a little while by swimming, but without electricity, there is no air conditioning or electric fans.

Fast Fact

There are more than 1,000 different Indigenous rainforest communities.

The rainforest also has a lot of bugs! Some of them are harmless, but others can be deadly. People who live and work in rainforests learn how to tell the

UNCONTACTED TRIBES

There are some Indigenous groups living in tropical rainforests, especially the Amazon, who have had little to no contact with the outside world. The largest number of these groups live in the border area of Peru, Brazil, and Bolivia. They are known as "uncontacted tribes." However, many tribes that are considered uncontacted were contacted at least once in the past and decided to move deeper into the rainforest.

Many tribes made this decision because they were treated poorly when they were first contacted. Outsiders are known for bringing disease, destroying the forest the Indigenous people rely on, and trying to change how the Indigenous people live—for example, by forcing them to follow a different religion, or belief system. Many uncontacted tribes defend their land with traps and weapons.

Life in the rainforest is not always easy, but to the people who call this beautiful place home, the hardships are worth it.

difference. The noise of frogs, bugs, birds, and other animals can get so loud that it drowns out conversations. Even at night, a rainforest is never silent.

Simple Pleasures

There are fewer things to do in a rainforest than there are in a city. People who live there often aren't able to go to movie theaters or plays, watch TV, play video games, or go shopping or out to eat. Instead, they spend time with friends and family. They may read, sing,

In a hot, wet climate, wood rots quickly. Indigenous people frequently need to build new houses as their old ones fall apart.

play musical instruments, or play games together. They also need to spend a lot of time working. Rainforest chores can include hunting, fishing, and clearing paths to walk on.

Fast Fact

There are no cities deep in the tropical rainforest, but there are a few on the edges of rainforests. These include Manaus, Brazil; Leticia, Colombia; Rurrenabaque, Bolivia; and Coca, Ecuador.

Ways of Life

There are many different groups of people living in rainforests around the world, and they all have different cultures, or ways of life. Some, such as the Sateré-Mawé people of Brazil, have been forced out of their traditional land and into cities. They no longer live the way their ancestors did, but they have worked to keep their culture alive.

Many Indigenous people have been hurt, killed, or driven off their land by non-Indigenous people. In some places, such as Australia, governments made it illegal for them to keep their culture alive. One of the few tribes to escape this fate is the Kuku-Yalanji people of North East Queensland, Australia. They are the only rainforest-dwelling tribe in Australia to keep their language, culture, and history alive. Their stories date back at least 50,000 years, to the earliest human settlements in what is now known as Australia.

Although all Indigenous groups have their own unique cultures, most share some things in common. They generally believe in living in harmony with nature. They take care of the land and the animals that live on it, don't take more from nature than they need, and see themselves as part of nature rather than seeing nature as something to be controlled and used. Many Indigenous groups are asking

Shown here is a group of Australian Aboriginal, or Indigenous, people performing a traditional dance.

non-Indigenous people to change their view of nature and do their part to keep the rainforests—along with everything and everyone living in them—safe for years to come.

Fast Fact

Samela Sateré-Mawé is a young environmental activist. She was inspired to take action by her culture's deep ties to the rainforest and by the climate activism of other young people.

THINK ABOUT IT!

1. Why do you think tropical rainforests are the best-known kind of rainforest?

2. What other animals can you think of that might live in each of the rainforest layers?

3. Why do so many plants we depend on grow in the rainforest?

4. Do you think Indigenous tribes who live in the rainforest should be contacted or left alone?

GLOSSARY

adapt: To change to fit conditions.

biodiverse: Having a lot of biological variety in an environment, as indicated by numbers of different species of plants and animals.

brackish: Slightly salty.

commercial: Relating to commerce, or the buying and selling of goods and services.

dense: Packed very closely together.

describe: To represent or give an account of in words.

dissolve: To mix completely into a liquid.

drought: A long period without rain.

ecosystem: A natural community of life that includes living and nonliving features.

nourish: To provide with food.

subcategory: A class or group that is part of a larger class or group.

thrive: To grow well.

traditional: Having to do with long-practiced customs.

unique: Special or one of a kind.

FIND OUT MORE

Books

Hayes, Vicki. *Surviving the Rainforest*. Minnetonka, MN: Kaleidoscope Publishing, 2021.

Hodge, Deborah. *West Coast Wild Rainforest*. Toronto, Ontario: Groundwood Books, 2024.

Markle, Sandra. *Secrets of the Lost City: A Scientific Adventure in the Honduran Rainforest*. Minneapolis, MN: Millbrook Press, 2022.

Websites

Kiddle: Rainforest Facts for Kids
kids.kiddle.co/Rainforest
Read more about the rainforest, and take a look at some cool photos.

***National Geographic Kids*: Rainforest Habitat**
kids.nationalgeographic.com/nature/habitats/article/rain-forest
Read more about the plants and animals that are found in rainforests around the world.

Rainforest Foundation US: Kids' Corner
rainforestfoundation.org/engage/kids
This website includes printable activities, such as word searches and mazes, as well as tips on how you can take action to help save the rainforest.

Publisher's note to educators and parents: Our editors have carefully reviewed these websites to ensure that they are suitable for students. Many websites change frequently, however, and we cannot guarantee that a site's future contents will continue to meet our high standards of quality and educational value. Be advised that students should be closely supervised whenever they access the internet.

INDEX